LORD BALFOUR
IN HIS RELATION
TO SCIENCE

LORD BALFOUR
IN HIS RELATION
TO SCIENCE

By

LORD RAYLEIGH

CAMBRIDGE
AT THE UNIVERSITY PRESS
1930

CAMBRIDGE UNIVERSITY PRESS
Cambridge, New York, Melbourne, Madrid, Cape Town,
Singapore, São Paulo, Delhi, Tokyo, Mexico City

Cambridge University Press
The Edinburgh Building, Cambridge CB2 8RU, UK

Published in the United States of America by Cambridge University Press, New York

www.cambridge.org
Information on this title: www.cambridge.org/9781107616448

First published 1930
First paperback edition 2011

A catalogue record for this publication is available from the British Library

ISBN 978-1-107-61644-8 Paperback

CONTENTS

PREFACE

It is the custom of the Royal Society to publish obituary notices of Fellows deceased. The present Memoir was written at the request of the Council of the Society, and appeared in the *Proceedings*. It has been thought, however, that a different circle of readers would be interested in it, and accordingly it is now republished.

Any attempt to summarise Lord Balfour's public career would have been out of place in the original publication, and is beyond my scope. This Memoir is limited to an account of his early history and mental development, his scientific and philosophical thought, and his administrative work for scientific, industrial, and medical research. In writing it I have been able to draw upon an intimate personal knowledge extending from my own earliest years, on the recollections of surviving relatives, and on the diaries of Balfour's sister, the Dowager Lady Rayleigh. In connection with Balfour's administrative work for science, valuable help has been given by Mr Thomas Jones, C.H., Captain A. F. Hemming,

PREFACE

C.B.E., Sir Frank Heath, G.B.E., Sir Walter Fletcher, K.B.E., and one or two other officials of the government departments concerned.

I may add that I have not drawn to any appreciable extent on the *Chapters of Autobiography* just published by Messrs Cassell. The present account of Balfour's early life is independent, and represents him and his surroundings as they appeared to other members of the family.

RAYLEIGH

Terling Place
Chelmsford
November, 1930

LORD BALFOUR
IN HIS RELATION
TO SCIENCE

ARTHUR JAMES BALFOUR was born in 1848.
He was the son of James Maitland Balfour of
Whittingehame, Haddingtonshire (died 1856),
and of Lady Blanche Gascoyne Cecil, second
daughter of the second Marquess of Salisbury.
The father was a country gentleman of fortune
and of some ability, but in no sense a philosopher
or a thinker, or even a reader. He served in
Parliament, organised a regiment of yeomanry
and was chairman of the North British Railway.
His career, however, was cut off at an early stage
by consumption. The mother, Lady Blanche
Balfour, came of an able stock. Although the
influence of the great Lord Burghley must be
considered to have been long since exhausted, the
abilities of that branch of the Cecil family had been
recently recuperated by two marriages: the first
with Lady Mary Amelia Hill, first Marchioness
of Salisbury; and the second with Frances Mary,
daughter of Bamber Gascoyne, and mother of

the third Marquess, Prime Minister to Queen Victoria, and of Lady Blanche Balfour. Lady Blanche shared to a marked extent her brother's ability. She directed the education of her children, and found means of stimulating their intellectual interests, with the greatest tact and judgment. She was particularly judicious in the choice of the books which she read to them. On occasion she would cut out passages which were tedious or unsuitable, and substitute connecting links rewritten by herself.

Her lessons in arithmetic are remembered for the clear way in which the subject was presented. The reasons for the operation of 'carrying', *e.g.*, were lucidly explained, in contrast to the usual dogmatic method.

Although Lady Blanche's own tastes were literary rather than scientific, she successfully directed the attention of her children to various branches of natural history, and, in the case of some of them, the interests thus aroused bore no inconsiderable fruit. Thus Gerald and Francis Balfour wrote an account of the local geology of East Lothian, which has permanent value as a contribution to the subject. Francis Balfour, it is hardly necessary to remind the reader,

[2]

eventually became Professor of Animal Morphology at Cambridge, a chair specially created for him. Those best qualified to judge held that his early death represented an irreparable loss to science.[1] His youngest sister, Alice Balfour, assisted him to some extent in his studies, and has ardently continued the pursuit of local entomology up to the present. Arthur Balfour's eldest sister had a distinct gift for mathematics and physics, and in later years was able to render valued assistance to her brother-in-law, Lord Rayleigh, during his tenure of the Cavendish Professorship of Experimental Physics at Cambridge. She appears as joint author of some of his papers on Absolute Electrical Measurements.

It will be seen then that there was a marked scientific tone in the family of which Balfour was a member. The same was traceable in his uncle, Lord Salisbury, who to some extent occupied his leisure with botanising, and with

[1] It has sometimes been imagined that Arthur Balfour derived his knowledge of science and his interest in it at second hand from his brother Frank. I am sure that no one with inside knowledge would share this view. The taste was innate in both of them, and came out in both, though in widely different forms.

experiments in electricity and magnetism, which he carried on in a private laboratory at Hatfield. Balfour seems to have taken less part than some of the others in these natural history studies. He had, for instance, no inclination for the hard work of a systematic search for fossils. But, no doubt, what was going on had its influence upon him.

In later life he took some practical interest in forestry on his estate, but he had no knowledge of horticulture, and no eye for bird life. On one occasion, well remembered in the family, he remarked that he had been disturbed by a bird which had flown into his study. His youngest sister, alive with the instincts of a naturalist, wished to know to what species it was to be referred. But the only description he could give was: "Oh, I don't know; average small bird". Although he had not this kind of knowledge himself, he respected it in others. As a young man, he had shooting, deer stalking and salmon fishing at his disposal, but he soon abandoned them in favour of lawn tennis and golf.[1]

[1] A fishing exploit at a very early age is remembered, however. He had succeeded in landing an eel, and was executing a dance in celebration of his victory with such

Balfour first went to school at the Rev. C. G. Chittenden's at Hoddesdon. His opinion of the school was expressed to the parents of another small boy in the words, "Send him to Chittenden's. It is the only place where I ever learnt anything". Mr Chittenden, when asked who was the ablest pupil he had had, replied "Arthur Balfour", and, although this judgment was given after his quondam pupil had made a public reputation, no doubt he had been of the same opinion throughout. Mr Chittenden was a man of wide general information, and an interesting talker, though a stern disciplinarian in school hours. Master and pupil seem to have had a warm regard for one another, and after the latter had left for Eton, it was Mr Chittenden's favourite relaxation to visit him there. While at the school, Mr Chittenden often took him out for walks. An interest in music was one that they had in common, and it is probable too that they discussed scientific topics, with which Mr Chittenden had some acquaintance. It is certain at least that Balfour dipped into various scientific

vigour that he danced into a bed of stinging nettles. His yells of triumph soon gave place to yells attributable to a very different emotion.

[5]

subjects at this time, though perhaps no more deeply than many boys do. The present writer possesses a copy of *Carpenter on the Microscope*, with the inscription "A. J. Balfour, April, 1860, Eton College".[1] Balfour is remembered to have brought home a frictional electrical machine, and to have made some attempts, though apparently without special tenacity or success, to carry out experiments in electrostatics, with home-made accessories.

As we have seen, he had impressed Mr Chittenden, and at Eton he similarly impressed William Johnson,[2] who seems to have shown more discernment than most of Balfour's masters or contemporaries at this stage. His intellectual development was not precocious, but, in the event, it went on much longer than that of some of his early friends, who had, for a time, seemed to stand on a level with him. Indeed, it may be said that he went on developing almost to the end.

It has sometimes been thought that Balfour, like his uncle, Salisbury, found the atmosphere

---

[1] *Sic.* It is stated in his autobiography that he went to Eton in September 1861.
[2] Later known as Cory.

of Eton uncongenial. I am sure that he never said anything of the kind in my hearing: on the contrary he pressed strongly for some of his nephews to be sent there with the words "Much the best school". Nor was he disposed at all definitely to condemn the classical system of school education, though he was himself included in that large majority who, after spending years under the system, fail to acquire a working knowledge of the classical languages.

During the Lancashire cotton famine of 1862–3, Lady Blanche Balfour conceived the idea of saving money for the help of the distressed artisans, and incidentally providing a valuable practical experience for her young family by domestic economies. A projected trip to the continent was abandoned, and the household at Whittingehame was much reduced, the family helping with the housework. During the summer holidays Arthur Balfour and his brothers made the beds and blacked the boots, while his sisters did the cooking.

The time approached for him to go to the university. He had not made much progress in, or shown aptitude for, mathematics and it was necessary to make up for lost time. His sister,

Eleanor (Mrs Henry Sidgwick), remembers reading elementary trigonometry with him for the 'Little-go'. They had to make it out from a book as best they could without the help of a tutor, getting up early in the morning, and fortifying themselves with bread and milk for the effort. When they came to the point that $\sin\theta/\theta$ has the limiting value unity as θ is indefinitely diminished, Arthur Balfour was dissatisfied with the demonstration in the book (probably not without reason according to modern standards of mathematical rigour). But he thought the point would be interesting if one could fully understand it.

In later life he was deeply interested in the philosophical foundations of mathematics, particularly in connection with the theory of probability, and deplored that he had not the technical knowledge to follow current developments either in that direction or in mathematical physics. He often regretfully commented, "I expect it is too mathematical for me".

Within the family circle, he took his place as the leader in intellectual interests. He it was who usually discovered to the others new avenues in literature. For instance, he came back from

[8]

school on one occasion brimming over with interest in 'Goēthe'.

The social tact which distinguished him in later life was innate, and already apparent in his boyhood. He was somewhat careless in the matter of dress, and very much detached from the smaller anxieties of everyday life. For example, he was driving with his eldest sister to a dinner. She expressed doubts as to whether the coachman was going the right way. "That", said Balfour, "is his affair."

He went up to Trinity College, Cambridge, in the October term of 1866, as a fellow-commoner, according to the custom of those days, as his father had done before him. This gave him the doubtful privilege of wearing a gown embroidered with silver,[1] and the valued one of sitting at the High Table with the dons. Here he was brought into contact with Henry Sidgwick and John Strutt (afterwards Lord Rayleigh), who were a few years senior to himself and fellows of the college. With them he formed an enduring friendship, which in each case developed into something more. For the former eventually

[1] He is said to have been the last, or almost the last, fellow-commoner.

married his eldest sister, Eleanor (1876), and the latter his second sister, Evelyn (1871).

It does not appear that Arthur Balfour impressed his individuality very strongly on the High Table at Trinity. Rayleigh remembered a discussion there, a few years later, as to which of the Balfour brothers had the most ability. Some were for Gerald, others for Frank. When he himself put in a claim for Arthur, the general opinion seemed to be that he was propounding a paradox. There were no doubt substantial reasons for awarding the palm to his younger brothers at that time. They had achieved high academic success, whereas Arthur Balfour did not rise above the level of a second class in moral science. His tutor, Henry Sidgwick, who, with Rayleigh and Rayleigh's younger brother, Charles Strutt, had formed the highest opinion of him, was disappointed, but not altogether surprised by this result. Balfour was also somewhat disappointed himself, though academic success had not been a prominent aim in his mind. The explanation seems to have been that he had paid too much attention to the current problems of philosophy, and not enough to its literature and history.

The truth is that his was a mind which could ill submit to the bondage of following a prescribed course of study. With him the motive must be his own personal interest, not the fulfilment of a task. That was and remained repugnant to him whether an academic text-book or the text of a parliamentary Bill was to be assimilated. In later life he several times gave public expression to his dislike of the tyranny of the examination system, though he did not pretend that it could be avoided.

Frank Balfour, who came up to Cambridge in 1870, took on the college rooms (A 4, New Court) which had been occupied by his elder brother, and was thus served by the same bedmaker. Adam Sedgwick, the zoologist, who was a devoted disciple of Frank Balfour, was fond of quoting her observation that "Mr Arthur Balfour left a great many books about, but Mr Frank read them through". Arthur Balfour had, in fact, an extraordinary faculty for getting hold of the essentials of a subject without apparently feeling the need for systematic study. Later in life when casually asked how long he could continue reading a stiff book, he put the limit, rather paradoxically perhaps, at ten minutes.

The following remarks[1] express his own point of view:

We misuse the word 'superficiality,' I think; sadly misuse it. Superficiality does not depend upon the amount of knowledge acquired. It is a quality rather of the learner than of the thing learned. The smallest amount of knowledge may be thorough in the sense in which the word should be used. Knowledge of the general principle may be obtained by those who have neither the time nor the ability to master the details of any particular branch of science; but to say that that smaller modicum of knowledge is therefore superficial, and therefore useless, is wholly to mistake what superficial knowledge consists in, and what education aims at. You may know very little, and not be superficial; you may know a great deal, and be thoroughly superficial. Superficiality is a quality of yourselves, not of the knowledge you acquire.

It was at about this stage of his career that he read Darwin's *Origin of Species*, and its effect on his point of view was profound. His own mentality, it is true, was in many ways very different

[1] Speech at the opening of New Hall of Battersea Polytechnic, February 3, 1899 (*The Times*). Reprinted in *Arthur James Balfour as Philosopher and Thinker*, selection by Wm. Short, 1912.

from that of Darwin. He had no store of detailed systematic knowledge, and it would probably not have been congenial to him to acquire it, or himself to attempt to sift the wheat from the chaff. But any one who reads the earlier chapters of *The Foundations of Belief* will see how much he had been influenced by the study of Darwin. Such reserve as he had was not founded on detailed criticism of Darwin's facts or methods of reasoning. The following quotation[1] will illustrate its nature:

It is wrong to suppose that these supreme values [*i.e.* what is highest and rarest] in æsthetic ethics and thought seriously count in the struggle for existence. Saints, philosophers and artists have never, so far as I know, been specially successful in rearing large families themselves; nor have they enabled the communities which admired and occasionally produced them to crowd out rival populations from the rich places of the earth. As Nature measures utility, they are useless. In no effective fashion do they make for survival. They are but casual excrescences on the evolutionary process forming no part of its essential texture. They are, on the naturalistic hypothesis, an accident of an accident.

Few things on the spiritual side of evolution

[1] *Theism and Thought*, p. 28.

are more interesting than this. It is not perhaps strange that the onward momentum of those developments which make for biological success should carry them into regions where all, or almost all, their survival efficiency vanishes away. But surely it is strange that they or something of them should acquire new and higher values which naturalism can hardly explain and certainly cannot justify.

Balfour throughout his life had the highest admiration for Darwin, "because", he said, "he was not a partisan—he really wanted to find out the truth—an attitude of mind seldom found among men of science, and never among theologians".

This opinion was not formed without the knowledge that comes of personal contact. He had been introduced to Darwin's home at Downe by the latter's son, George, who was one of his early friends at Cambridge, and to whom he remained warmly attached to the end.

Shortly after taking his degree, Balfour went with his friend, John Strutt, to visit the Gladstones at Hawarden Castle. Rayleigh often referred to an incident on this visit which evidently produced a strong impression upon him. For some reason it was necessary to ease the labours of the household, and in consequence Mr Gladstone took Balfour and Strutt to dine at the village inn. Probably the two former took the chief part in conversation, and Strutt, it is likely, was a comparatively silent onlooker; be that as it may, he presently perceived that Balfour was not taking the veteran statesman seriously, but was amusing himself with a psychological study, and, as he often related the incident afterwards, "was playing Gladstone like a fish". Balfour's age at this time, be it remembered, was only twenty-two years.

The circumstances which led to Balfour's entering on a political career are explained in the fragment of autobiography which has appeared recently. We are here rather concerned with the question of why he did not take up science. The following extract from the journal of his sister Evelyn (Lady Raleigh) of September 10, 1888, will explain his own point of view:

Yesterday *A.* was as usual expressing a wish that he had been a scientific man instead of a politician.

E. Why did you not devote yourself to science then?

A. I did not consider I was capable of it.

E. John[1] does not agree with that, he says a conversation on science with you sometimes strains his powers.

A. Oh, nonsense—but I was too lazy, I should never have had patience for the drudgery.

It is perhaps scarcely worth while to devote much consideration to what *might* have been. Other observers will probably agree with Balfour's own view that attention to drudgery was not his strong point. At the same time it is difficult to believe that he can have got through so many years of successful public life without having done a good deal of it.

He was certainly not an enthusiastic politician. He remarked on one occasion (1893) that his mind did not naturally run to politics. He never thought of them in bed, which was the test. He regarded them with a calm interest, but as for getting excited over them as some people did— he could not do it.

He was, I think, comparatively indifferent to both the trials and the rewards of official life as a party politician. He did not often trouble to inform himself as to what people were saying about him, and he seldom looked at a newspaper.

[1] Lord Rayleigh.

He remarked that everything conceivable had been said of him, good, bad and indifferent, and that in view of this it was difficult either to be much depressed or much elated by the sum total of the opinions expressed. He may have been somewhat hurt at having temporarily lost the confidence of the Unionist Party in 1911, but if this may be suspected it is not from anything that could be seen at the time, but from slight indications elicited by later events.

His detached attitude towards criticism may perhaps have made the sheltered life of a student relatively less attractive to him than to many others equally well qualified to pursue it. As to the honours and rewards of a public career, I know on the best authority possible that he accepted most of them with reluctance, and because circumstances made it difficult to do otherwise.

Finally, it may be remarked that most people realise more clearly the disadvantages of the career they have actually adopted than of the one they might have adopted and did not.

Balfour was first elected to Parliament as member for Hertford in 1874, and from this time onwards became increasingly immersed in public activities which do not fall within the scope of this account; at no time, however, were his intellectual interests allowed to fall altogether into the background, and his first speculative work, *A Defence of Philosophic Doubt*, appeared in 1879, shortly after his return from the Berlin Conference of 1878.

The title was not chosen without careful consideration, but none the less its meaning was widely misunderstood. As anyone who looks even casually into the book may see, doubt about the views of Mill and Spencer is advocated, not doubt about popular theology. "It appears", he says, "that the practical conclusions I draw from a sceptical philosophy have little or no tendency to alter the internal structure of any actual or possible creed."

The *Defence of Philosophic Doubt* discusses in turn each of the various theories of knowledge which at the time of writing could be considered to have any important following. J. S. Mill, Spencer, Kant, Hamilton, and the agnostic scientific school represented by writers like Huxley and Leslie Stephen, are each subjected in turn to a critical analysis, which is carefully limited to essentials. The author is never led into controversial bypaths, however tempting, in order to score a point. He concludes that not one of these schools of thought is self-consistent, though he does not claim to be able to produce anything better. At the same time he carefully explains that he, like everyone else, cannot help accepting in practice the methods and conclusions of science, in spite of the incoherence he finds in them, regarded as a logical system. As a 'practical result' he recommends that scientific conclusions should be provisionally adopted alongside of theological ones, even at the cost of apparent inconsistency. I do not know of any evidence that other thinkers at that time found themselves able specifically to accept this recommendation. It must be admitted, however, that an attitude very like this has been taken up in modern science, in using the wave theory of light to co-ordinate one set of phenomena, and the corpuscular theory to co-ordinate another set.

The general line of argument in this book has much in common with Balfour's second and better-known book, *The Foundations of Belief*, on which he placed, perhaps, a higher value.

About the time of publication (December 1894) he said, in intimate conversation, that he felt he had a message to give, which he was trying to express in this book, and which was of far greater importance than anything he had done or could do in politics. He was especially pleased and encouraged to find that his brother-in-law, Henry Sidgwick, though not always in agreement, thought highly of it; and as the public were anxious to have the views of a conspicuous public man on questions of such fundamental interest, the book sold largely.

In this, as in the author's other philosophical works, the greater part of the text is occupied with argument destructive of the philosophical point of view which he refers to as Naturalism, and, so far as I have been able to gather, it is this sceptical criticism, carried out with a wealth of illustration from the scientific field, and an easy dialectical mastery, which leaves the most marked impression on the minds of the generality of readers. As a brief example of the method we may quote[1]:

Though we are quite familiar with the fact that illusions are possible, and that mistakes will

[1] *The Foundations of Belief*, 8th ed. p. 118.

[20]

occur in the simplest observation, yet we can hardly avoid being struck by the incongruity of a scheme of belief whose premises are wholly derived from witnesses [the physiological mechanism of the senses] admittedly untrustworthy, yet which is unable to supply any criterion, other than the evidence of those witnesses themselves, by which the character of their evidence can in any given case be determined.

The author, however, was most anxious that the constructive part of his argument should not be overlooked, or misunderstood:

I seem [he wrote[1]] to have given certain of my critics the impression that the principal, if not the sole, object of this work was to show that our beliefs concerning the material world and those concerning the spiritual world are equally poverty stricken in the matter of philosophic proof, equally embarrassed by philosophic difficulties. This, however, is not so.... The dissipation of a prejudice, however fundamental, can at best be but an indirect contribution to the work of philosophic construction. Concede the full claims of the argument just referred to, yet it amounts to no more than this—that while it *is* irrational to adopt the procedure of naturalism, and elevate scientific methods and conclusions

[1] *The Foundations of Belief.* Introduction to 8th ed. pp. xvii, xviii, 1901.

[21]

into the test of universal truth, it is *not* necessarily irrational for those who accept the general
methods and conclusions of science to accept
also ethical and theological beliefs which cannot
be reached by these methods, and which, it may
be, harmonise but imperfectly with these conclusions. This is indeed no unimportant result;
yet, if the argument stopped here, it might not
be untrue, though it would assuredly be mis-
leading, to say that the following essay only con-
tributed to belief in one department of thought
by suggesting doubt in another. But the argument does not stop here. The most important
part has still to be noted—that in which an en-
deavour is made to show that science, ethics,
and (in its degree) æsthetics, are severally and
collectively more intelligible, better fitted to
form parts of a rational and coherent whole,
when they are framed in a theological setting
than when they are framed in one which is
purely naturalistic.

During the interval between his retirement
from the leadership of the Unionist Party in
November 1911 till the outbreak of war, Balfour
had, perhaps, more leisure for intellectual pursuits than at any other period of his mature life,
and a near observer described him as like a bird
which had escaped from a cage. To this period

belongs the photograph reproduced herewith,[1] the holiday attire in which he appears harmonising with his mood at the time. This interval of comparative leisure was partly employed in preparing the first series of Gifford Lectures delivered at the University of Glasgow in January and February 1914, and published under the title *Theism and Humanism*. The general scope and mode of treatment is not unlike that adopted in *The Foundations of Belief*. Their success was extraordinary; the audiences amounted to something like two thousand and increased beyond the limit of seating accommodation as the course of ten lectures proceeded. He did his best to avoid technical language, which he considered was very apt to mask a confusion of thought; it was impossible however to avoid words like *empirical* and *a priori*. Two ladies were heard discussing who this *a priori* might be—some Italian philosopher they supposed! However, in spite of rather discouraging symptoms of this kind, he hoped and believed that most of his audience carried away something.

The services which Balfour's party had so

[1] Taken by Admiral Strutt in the conservatory at Terling.

easily dispensed with in times of peace were
found necessary by his country in the hour of
stress, and he was called to be First Lord of the
Admiralty in May 1915, shortly before the lectures were actually published. His brief interval
of comparative leisure was at an end.

The second course of Gifford Lectures was
necessarily deferred till after the war. They were
delivered in 1922–3 and published under the
title *Theism and Thought*.

Balfour, like many of his relatives, felt a
sympathetic interest in psychical research. Some
of his political followers were disposed to com-
plain of this. They classed it with Bi-metallism
and Female Suffrage, and considered that all
these 'fads' injured his position as a leader.
When criticism of this kind came round to him,
he said that he was not prepared to give up his
'fads', and that if a choice was necessary he
would sooner abandon politics.

In 1894 he gave a Presidential Address to the
Society for Psychical Research[1]. In this, among
other topics, he emphasised the warning to be
taken from the incredulous attitude of the
scientific world towards hypnotism:

[1] Reprinted in *Essays, Speculative and Political*, 1920.

[24]

There were, indeed, a good many doctors and other men of science who could not refuse the evidence of their senses, and who loudly testified to the truth, the interest, and the importance of the phenomena which they witnessed. But if you take the opinion of men of science generally, you will be driven to the conclusion that they either denied facts which were obviously true, or that they thrust them aside without condescending to submit them to serious investigation.

Balfour was, I believe, convinced of the reality of telepathy, and his conviction was fortified by some casual experiments with Prof. Gilbert Murray in which he took a personal part, and which excited a good deal of attention in the newspapers. On the alleged physical phenomena of spiritualism he had an open mind. He was, for example, unable lightly to dismiss the concurrent testimony of the late Lord Crawford and the late Lord Dunraven to some of the most marvellous happenings. He had known these men, and respected their capacity and good sense. Pressed to sum up, he answered, in a truly scientific spirit, that more experiments were needed.

As I have endeavoured to show, Balfour's

knowledge of the essentials, rather than of the details of contemporary science, was wide. When he was called upon to speak publicly on scientific or semi-scientific questions, he was usually able to illustrate a topic or to choose one from his own knowledge. Thus, when he was President of the British Association at Cambridge in 1904, he delivered an address in which what were then novel views of the electronic constitution of matter were discussed in their philosophical aspect. This was perhaps the first occasion when emphasis was laid before a popular audience on the glaring discrepancy between the new ideas of the atom, with its relatively vast inter-electronic spaces, and the old philosophic distinction which made shape a 'primary' property of matter, existing independent of the observer, while secondary qualities, such as colour, were thought to have no such independence. It is not, I think, an exaggeration to say that these conceptions are found novel and illuminating by many educated people even now, more than a quarter of a century after the address was delivered.

Foreign savants who attended as guests of the Association were a good deal astonished at the range of scientific knowledge of the speaker, and some of them were disposed for a moment to doubt whether he could really be the Prime Minister, in part because he wore no decorations. After the opening meeting Balfour attended some of the sectional proceedings, in which he was able to take part. No doubt at times he used the arts of the experienced public man in making a necessarily limited knowledge go as far as possible. But he was always eager to learn more.

An incident remains in my mind as an illustration of this: it was, I think, during the autumn of 1928. We had been out for a country walk together, perhaps almost for the last time, and after the conversation had ranged over a great variety of topics including the morals of the present generation (which he did not think really worse than those which prevailed during his youth), the distribution of honours, the prospects of future taxation, and the literature of the eighteenth century, it came round, I do not remember exactly how, to the electromagnetic theory of light. "There are things I sometimes talk about", he said, "which I find it very difficult to get any grasp of. I think I understand pretty well the relations between electric currents and magnetism, but I cannot really form any conception of Maxwell's theory of the propagation of electromagnetic waves." I said I thought it was too difficult for general treatment. Maxwell's calculations must be followed through in order to get any insight into it.

He declined, however, to be put off in this way, and insisted on my trying to expound it. I did what I could, helping myself out with diagrams drawn on the road with a walking stick. However poor the attempt, he seemed fascinated with the subject, and I remember trying to think of anyone else who would have been able to learn anything from such an explanation.

Relativity, occupying as it does the borderland between science and philosophy, interested him deeply. One remark he made on this subject is of some psychological interest, whether one is able to agree with it or not. He complained that the popular expositors put an undue and unnecessary stumbling-block in the way of their readers when they traced out the paradoxical consequences of extreme suppositions, such as observers travelling on projectiles which moved with the velocity of light. I tried to plead that in science hard cases made *good* law, but he was not to be moved from his position. As regards the technique of physical measurements he was not so much interested. He could, of course, appreciate well enough the necessity of reaching a certain standard of precision in order to resolve a particular problem in hand; for instance, the standardisation of machine parts to go together without fitting. But he was not prepared to regard improved accuracy of measurement as an end in itself. This was evident enough to those who had the opportunity of discussing scientific subjects with him. Others will find an illustration of it in a reference to the measurement of solar parallax by the transit of Venus in one of his essays.[1]

Balfour was elected into the Royal Society under Statute 12 as early as 1888, and served on the Council in 1907–8 and again in 1912–14. In 1920 when the question of an election to the Presidency came up, the retiring President, Sir J. J. Thomson, was commissioned by the Council to find out whether Balfour would allow his name to be put forward, on the understanding that the Society would want him as an active, and not merely as an ornamental, president. Balfour

[1] *Essays and Addresses*, 3rd ed. 1905, pp. 23–24.

was more than commonly pleased to receive so great a mark of the confidence of the scientific world; but he was already President of the British Academy; as Lord President he was responsible for the Department of Scientific and Industrial Research; he was taking a very leading part in the affairs of the League of Nations; and, in addition to all this, the Cabinet was meeting once, twice, or even three times a day. He felt that to add to this programme of work was impossible.

We have seen how Balfour had acquired a far wider scientific culture than usually falls to the lot of practical politicians. This culture had helped his philosophical studies, and had enabled him with the greater force to give public expression to his belief in the national importance of scientific pursuits. But otherwise it had not had much constructive outcome; nor can it have seemed likely in 1919 that at the age of seventy-one years it would ever do so. His political career seemed for the moment to be closing, and he was determined not to do any more heavy political work of the ordinary kind— leading the House of Lords, for example. He had always been inclined to rate rather low the value of what could be accomplished in that way[1] and he felt that he had earned his rest from it. Yet his public career was about to enter on a new phase, in which his scientific interests and acquirements were to bear no inconsiderable fruit. This phase seemed to himself more interesting and at least as important as any that had preceded it.

It is necessary to go back somewhat in time and to recall that in connection with Mr Lloyd George's National Health Insurance Act of 1911, a large scheme of medical benefit was supplemented[2] by a considerable provision for medical research. Balfour spoke strongly in favour of this in the House of Commons, but the decision had apparently been taken already, and I have not been able to learn that he was definitely instrumental in bringing it about. The Medical

[1] Extract from Lady Rayleigh's journal, June 16, 1892: "Paderewski was at the Royal Society Soirée last night, and in discussing it A. remarked of the scientific guests, 'They are the people who are changing the world and they don't know it. Politicians are but the fly on the wheel—the men of science are the motive power'". The same point of view is more elaborately set out in his essay, "A Fragment on Progress", republished in *Essays and Addresses*, 1905. See particularly pp. 260–2.

[2] Probably on the initiative of the late Sir Robert Morant, K.C.B., of the Education Office.

Research Committee was the outcome in 1913, and this in 1919 became the Medical Research Council, under a Privy Council Committee and the Lord President. The war led to a great stimulus of State-supported science; for scientific problems arose in connection with almost every war activity. One may recall aeronautics, radio communications, anti-submarine work, poison gases, medical and surgical war problems such as antitetanus inoculation and plastic surgery, sound ranging—but the list is almost inexhaustible. In this way politicians and administrators were brought into contact with science and with scientific men. The ice was broken; budding scientific institutions underwent a forced growth; and when the war was over its lessons could not be unlearnt, and the growth which had taken place proved on the whole to be permanent. It is a significant fact that when the post-war *régime* of retrenchment set in, it was not extended to research activities.

The Department of Scientific and Industrial Research was founded in 1915 and placed under the Lord President of the Council. Many new State scientific activities have been placed under the Department—fuel research, low temperature

research, building research, forest products research, chemical research, radio research, and
two long-established scientific institutions—the
Geological Survey and the National Physical
Laboratory—were also annexed to it.[1] The Department has many other responsibilities besides
these. It administers large funds for research
purposes in academic institutions and industrial
research associations.

Balfour, after his return from Paris in 1919,
retired from the Foreign Office, in accordance
with his determination to do no more heavy
political routine, and undertook the lighter duties
of Lord President. He then found himself in
possession of the heritage which has just been
sketched in outline. He probably had little up-
to-date information about his new responsibilities, for since 1915 his time had been fully
occupied at the Admiralty and the Foreign
Office, and he cannot have been able to give

[1] The latter had been founded in 1900. Balfour was
then First Lord of the Treasury, and was most sympathetic to the scheme. Indeed the suggestion that the
laboratory should go to Bushy Park originated with him.
War requirements had increased the extent of the laboratory's operations out of all knowledge, and it had become
impracticable for the Royal Society to continue to carry
the liability for balancing income and expenditure.

much attention to anything else. One evening in the autumn of 1919 the Secretary of the Medical Research Council, who was somewhat pressed with work, was told that there was a gentleman to see him—Mr Balfour. "I did not make any appointment, did I?" he said. "I do not think I can see him. What Mr Balfour is it?" To his astonishment it turned out to be the Lord President. It was quite a new thing in his experience that a Minister should call on an official who served under him, and he had no reason to think that previous Lord Presidents had even known where the office was. Balfour opened the conversation by remarking that his new appointment seemed to give him very little to do, but that he was delighted to find that he was the head of two research organisations. He had come to learn whether there was anything he could do to help.

It soon appeared in practice that in those troublous times other ministerial duties and distractions made far greater claims than he had anticipated, and during the period of office from 1919 to 1922, when he went out at the fall of the Coalition Government, he was not able to attend the meetings of the Medical Research Council.

As responsible Minister he made no trouble about details. When formal matters were put before him, he was content with the assurance of his advisers that all was right, and initialled them without further comment. "And now", he would say, "tell me something interesting."

For a short interval between 1922 and 1924 Balfour's connection with the Council was intermitted.

Lord Curzon became Lord President in 1924 and when Lord Irwin, who had been chairman of the Medical Research Council, went to India in 1925, Curzon sent to ask the advice of Balfour, who was not then in office, as to who should succeed him. It soon appeared that he was willing to serve himself, and so it was arranged. He became chairman, and attended every meeting. If he had a fault, it was not one that would be expected in a political chairman. He was if anything too prone to spend time at the Council in discussing the scientific rather than the administrative aspects.

On Lord Curzon's death in 1925, Balfour again became Lord President, at Mr Baldwin's special request. The problem then presented itself whether he could continue as chairman of the advisory body, whose duty it was to give advice to himself.

The Council did not wish to lose him and arguments were readily found to justify the course which everybody desired should be taken. In the event he remained chairman until his last illness in 1930.

Balfour followed the developments of medical research with keen interest. He had known in his youth Sir James Simpson, of Edinburgh, the first to use chloroform as an anæsthetic, who was an intimate friend of Lady Blanche Balfour. He had early become acquainted with the methods of preventive inoculation by killed bacterial cultures by Sir Almroth Wright, to which he had become a firm convert. During the most active part of his Parliamentary career he was frequently placed *hors de combat* by feverish colds. It was often thought that these illnesses were merely diplomatic, but the suspicion was unfounded. He tried preventive inoculation, and benefited greatly from it. In this way he was brought into contact with the leading worker on the subject and his scientific interest was aroused. His friends and relatives heard a great deal at that time about streptococci and staphylococci.

On the question of ordinary vaccination against smallpox his views were somewhat heterodox. In the autumn of 1898 he said in conversation with an intimate circle that he had been in favour of inserting a conscience clause into the Vaccination Act recently passed, and that since the Bill had passed he had looked somewhat further into the evidence, with the result that he had grave doubts whether vaccination was of any use at all. It was, he said, a question of statistics, not of science. In 1802, when vaccination was introduced, smallpox had decreased enormously, but much of the decrease had taken place, it was found, before vaccination became general. In 1876, when the belief in vaccination was at its height, we had an epidemic of smallpox which threw into the shade the worst epidemics of the eighteenth century. There were, he said, other facts of a similar character.[1]

During Balfour's chairmanship of the Medical Research Council he did not fail to attend the occasional afternoon teas at the Council's Institute for Medical Research at Hampstead, when

[1] These remarks are from a contemporary record. The present writer is not qualified to offer any comment upon them.

workers such as Gye and Barnard gave an account of the problems on which they were engaged. He took every opportunity of pointing out in occasional speeches the debt which the nation owes to medical research workers and the danger of allotting too large a share of the nation's gratitude and support to those who apply medical knowledge, to the neglect of those who originate it in the seclusion of the laboratory, far away from the public eye.

The Department of Scientific and Industrial Research benefited not less than the Medical Research Council from Balfour's connection with it, which likewise began in the autumn of 1919. Here also he was in the habit of attending Council meetings, sitting next to the chairman, Sir William M°Cormick, with whom he was on friendly terms. Indeed, friendliness was always the note in his relationship with Civil Servants, whether high or low. Thus, after official interviews at Balfour's house in Carlton Gardens, the secretary of the Department was commonly asked to stay to lunch, and the subject of discussion was often carried further in Balfour's family circle. The secretary's secretary was known to remark that during her official experience she had only twice been personally addressed by a Cabinet Minister. The first time by one who shall be nameless, and who said, "You may go"; the second time by Balfour, who said, "Pray don't go, pray don't go".

To return however to the meetings of the Advisory Council. Balfour's contributions were mainly in the form of questions, which often opened a new point of view to the Council, and always referred to essentials.

His policy did not always err on the side of caution, and he was on occasion prepared to go beyond the advice of his Council. Thus the fuel research board which is under the Department had been concerned with the hydrogenation of coal. There was the opportunity of purchasing for £35,000 ten years of experience on the problem, a half-scale plant, and all necessary opportunities for acquiring the knowledge to work it. The Council were not enthusiastic, some of them who were best qualified to express an opinion thought that the possibility of practical results was remote, and that the money from the limited budget would go further in other ways. However, a resolution of cold approval was eventually passed, and Balfour decided that the possibilities were so important from a national point of view that the expenditure was a legitimate gamble.

Again, when difficulty was anticipated in financing the promising work of the magnetic laboratory at Cambridge, under Dr Kapitza, rather than that the work should be hindered, he was prepared to shoulder the responsibility of financing it from the Department. This might be held to be somewhat irregular, but he expressed his willingness to defend it in Parliament. Had he been called upon to do so, his dialectical resources would, no doubt, have proved quite adequate to the occasion.

Balfour's attendance at the Council meetings, and his afternoon calls—unannounced and without appointment—at the secretary's office, were found most stimulating and encouraging. On the latter occasions, he would sit in an armchair discussing difficulties and hopes of achievement. He would ask what the truth really was about matters which were less ripe. Often the director of fuel research and other technical officers would be called in to the discussion, and the same sense of stimulus was experienced by them.

During Lord Curzon's term of office as Lord

President, the question had been mooted of devising machinery for co-ordinating the work of different government departments, and bringing the activities of the research organisations now grouped under the Privy Council into more effective relation with them. The initiative came from members of the Civil Service. Lord Curzon's attitude was not very sympathetic at first, and further discussion was cut short by what proved to be his last illness. Balfour, as chairman of the Medical Research Council, was conversant, as Mr Baldwin well knew, with these preliminary moves, and when he succeeded Curzon in office as Lord President (1925), Mr Baldwin commissioned him specially to take up the problem.

One suggestion was to appoint a standing committee of leading scientific men. Balfour said: "No. In the first place you are putting an undue burden on them. Secondly, you will have to make an invidious choice as to whom you ask to join the committee, and any man you select will be adapted perhaps to one problem, and not at all adapted to another. It would be far better to have a more elastic system and to imitate the organisation of the Committee of Imperial

Defence ".[1] Distinguished Civil Servants who had been called into consultation were soon won over to see the wisdom of this point of view.

The Committee of Imperial Defence, the reader may be reminded, was of Balfour's own contrivance. He devised it in 1904 when he was Prime Minister, because he found that no attempt was being made by the departments to pool their information, or to arrive at clear mutual understanding about war problems that might affect more than one of them. Thus the Navy said that *they* made us safe from invasion, while the Army said that we were in the greatest danger because *they* were not maintained at a sufficient strength; and the Post Office had no idea that *they* had anything special to do in war,

[1] It is right to state that the idea of imitating Balfour's Committee of Imperial Defence for civil purposes was originally mooted by Lord Haldane as early as 1918.

The Labour Government of 1924, of which he [Haldane] was a member, considered the question further, apparently with economic rather than scientific enquiries in view; but nothing was done before they left office in the autumn of 1924. After the Labour Party returned to office in 1929, the organisation was modified in the way that Balfour had deprecated, and a standing committee was set up with the title of Economic Advisory Council. Whether this change will prove advantageous or permanent the future alone can show.

[42]

though in fact the part they have to play is very important. And similarly in other cases.

The organisation originally set up by Balfour in 1904 to remedy this state of things is a committee, with only one permanent member—the Prime Minister. The other members are such persons as he may summon to sit—mainly his colleagues in the Cabinet, but also the heads of the fighting services, and occasionally other experts. But this is only the first stage. The detailed investigations are carried out by sub-committees in which a much larger latitude of choice is exercised, and there is as the backbone of the whole structure a permanent secretarial staff which serves the main committee and the sub-committees.

The above may seem somewhat of a digression, but its relevance will soon appear. Balfour's scheme of 1925 was closely copied from his scheme of 1904, and in fact it was possible to make use of the same offices and the same secretariat that already served the Cabinet and the Committee of Imperial Defence. Mr Thomas Jones acted as principal secretary, with the assistance of Captain A. F. Hemming. The title chosen was the Committee of Civil Research.

Balfour was eager to make a beginning and

to put the machinery thus created to the proof. The first subject of investigation was the Tsetse fly disease of Eastern and Central Africa. This had bearings both on health and on agriculture, and concerned the Colonial Office, the Fighting Services, the Foreign Office, and the Dominions. It was therefore typical of the kind of problem which the committee was created to deal with.

Balfour (acting for the Prime Minister) presided at the first meeting, and, as anticipated, a great lack of co-ordination was revealed. Under the Foreign Office many thousands a year were being spent in the Soudan, and spent to good advantage, particularly in the direction of quarantine. The Colonial Office representative admitted, under Balfour's cross-examination, that the Administration of Uganda knew nothing of this money being spent, and that they themselves had refused a trifling sum in respect of this kind of research. The man who was fighting the Tsetse fly in Uganda did not even know the man who was officially working at the same problem in the next colony.

The committee has remained active up to the time of writing. The Governor of Tanganyika, who happened to be in England, was later summoned to the committee, and as a result was glad to acknowledge the enlightenment it had given him, and to make consequent changes in his financial policy; and there were many other similar cases.

Balfour's influence did much to orient the policy of the Colonial Office towards the scientific aspect. This was the result of genuine conviction which he was able to impart, and not merely deference to superior authority. The machinery which he set up for the first time gave scientific men direct access to Ministers, without the intervention of lay officials, who were often unsympathetic to the scientific point of view, and deprived scientific advice of much of its effect.

Other scientific or semi-scientific subjects considered by the Committee of Civil Research during Balfour's *régime* (1925–9) were mineral content of natural pastures, Severn barrage, quinine supplies, dietetics, British pharmacopœia, research co-ordination, Kenya native welfare, geophysical surveying, Great Barrier Reef, mechanical transport, irrigation research, radium supplies, locust control.

The inauguration of the Committee of Civil Research was perhaps the last important achievement of Balfour's long career. The concluding

[45]

phase of that career cannot be better summed up than in the words of one who was able to observe it at close quarters[1]:

The Lord Presidency used to be considered a general utility office. He converted it into a Ministry of Research. The idea was not born in his fertile brain, for a committee of the Privy Council for Scientific and Industrial Research and a similar committee for Medical Research had been established during the war, and Lord Haldane's committee on the machinery of government had recommended the creation of such a Ministry. But Lord Balfour it was who turned an experiment, which many thought destined to disappear with other war-time devices, into a reality which is now generally recognised as a permanent and essential part of modern government. His unparalleled prestige in the political and intellectual worlds, his liberation from the rough-and-tumble of party politics were favourable circumstances, but his abiding faith in the power of science to promote the happiness and well-being of man, his enthusiastic interest in the advance of knowledge, his sympathy with the scientific outlook and with young people, and his long experience of the way in which things have to be done in Great Britain, were the decisive factors.

[1] Sir Frank Heath, G.B.E., K.C.B.

Milton Keynes UK
Ingram Content Group UK Ltd.
UKHW041519181024
449640UK00009B/73